Earth Materials:
The Mystery Rocks

by Emily Sohn and Pamela Wright

Chief Content Consultant
Edward Rock
Associate Executive Director, National Science Teachers Association

Chicago, Illinois

Norwood House Press
PO Box 316598
Chicago, IL 60631

For information regarding Norwood House Press, please visit our website at www.norwoodhousepress.com or call 866-565-2900.

Special thanks to: Amanda Jones, Amy Karasick, Alanna Mertens, Terrence Young, Jr.

Editors: Barbara J. Foster, Diane Hinckley
Designer: Daniel M. Greene
Production Management: Victory Productions, Inc.

This book was manufactured as a paperback edition. If you are purchasing this book as a rebound hardcover or without any cover, the publisher and any licensors' rights are being violated.

Paperback ISBN: 978-1-60357-284-2

The Library of Congress has cataloged the original hardcover edition with the following call number: 2010044543

© 2011 by Norwood House Press. All Rights Reserved. No part of this book may be reproduced without written permission from the publisher.

Printed in Heshan City, Guangdong, China.
190P—082011.

CONTENTS

iScience Puzzle	6
Discover Activity	9
What Is Rock?	11
What Are the Three Types of Rock?	13
What Is the Rock Cycle?	24
Connecting to History	26
Science at Work	27
Solve the iScience Puzzle	28
Beyond the Puzzle	29
Glossary	30
Further Reading/Additional Notes	31
Index	32

Note to Caregivers:

Throughout this book, many questions are posed to the reader. Some are open-ended and ask what the reader thinks. Discuss these questions with your child and guide him or her in thinking through the possible answers and outcomes. There are also questions posed which have a specific answer. Encourage your child to read through the text to determine the correct answer. Most importantly, encourage answers grounded in reality while also allowing imaginations to soar. Information to help support you as you share the book with your child is provided in the back in the **Additional Notes** section.

Words that are **bolded** are defined in the glossary in the back of the book.

What Stories Do Rocks Tell?

Some **rocks** are flashy. They sparkle. They shine. They have streaks of color. They catch your eye. You might want to pick them up.

Many people collect rocks. Rocks are not only interesting to look at. They tell stories about the Earth.

In this book, you will learn some fun facts about rocks. You will also find clues to help you solve a puzzle. You need to name some mystery rocks!

iScience PUZZLE

Mystery Rocks

You are walking through the school's science lab when you find a box. Inside is a rock collection. There are lots of beautiful stones. Each one has a label that says what kind of rock it is. You are careful with the box when you pick it up. But then you trip over a chair. Three of the rocks lose their labels. Oh, no! How can you figure out which one is which?

an assortment of rocks

There are three loose labels in the box. One says "**basalt.**" One says "**marble.**" A third says "**conglomerate.**" You line up the three nameless rocks. You examine them one by one.

Rock 1:
This rock is dark gray and black. It has little holes in it. When you look closely, you see specks in the rock that look like salt or sand. These are called grains. This rock has both small grains and large grains.

Rock 2:

This rock is lightweight and bumpy. It is made up of round pebbles that look like they are glued together. There is even a **fossil** in the rock!

Rock 3:

Streaks run through this rock. The grains in it are large.

Can you put the right label on each mystery rock? Read on. By the end of the book, you should know which is which!

DISCOVER ACTIVITY

Sort the Rocks

Go outside to a park or yard. Gather some rocks. Try to pick rocks that don't look like one another. Go back inside and spread the rocks on a table. How are the rocks different from each other? How are they alike? In this activity, you will sort the rocks. You will use their **properties** to do this. A property is a way to describe an object. Color, shape, and feel are types of properties.

Look closely at your rocks. Learn as much as you can about them.

Take a close look at each of the rocks you collected. You may want to use a magnifying lens.

- What color is it?
- Are there holes in it?
- Can you see layers in it?
- Is it hard or soft?
- Is it shiny or dull?
- Can you see any grains in it?
- Pick up each rock. Does it feel heavy or light?
- How else can you describe the rock?

Now, sort the rocks. Start by picking a property, such as color. Put all brown rocks in one group. Put the red ones in another. And so on. Now pick another property. Group them again. How many ways can you find to group the rocks?

Remember the rock collection you found in the lab? Which of your groups would those three rocks fit into?

If you study rocks closely, you can discover similarities and differences among them.

What Is Rock?

You may be surprised to hear this. But that is a hard question to answer! Rocks come in many forms. There's one thing most rocks have in common. They are made of **minerals.** A mineral is a substance found on the Earth. It is made by processes that occur naturally on the Earth. It is not made from plants or animals. It is solid, and often has flat sides and sharp edges.

A sampling of nine different minerals

Scientists have identified thousands of minerals. Salt is a mineral. So are gold, copper, quartz, and gypsum.

different types of minerals

Rocks are made of minerals. They form over time. And they form in a variety of ways. Each type of rock has its own set of minerals. Experts look at the minerals in a rock. That tells them what kind of rock they are looking at.

Take another look at your mystery rocks. Can you see minerals in them? Are there just a few minerals? Or are there lots of minerals?

What Are the Three Types of Rock?

Look down toward your feet. Imagine you could see through the ground. Deep down in the earth, the planet is made of rock. If you go deep enough, the rock is a hot liquid.

Hot liquid rock from volcanoes hardens to form new rock.

How do you think the tree roots might change this rock?

Boom! A volcano erupts. Liquid rock comes to the Earth's surface. It hardens and becomes new rock. This is just one way that new rocks form. Sometimes, pebbles and sand get squeezed together. This makes new rocks, too. Other times, old rocks change into new rocks.

How do you think your mystery rocks formed?

a geologist studying rocks

Geologists are scientists who study the Earth. They talk about three main groups of rocks. These are **igneous, sedimentary,** and **metamorphic rocks.** Each group forms in a different way.

Do you think your mystery rocks all belong to the same group?

It's possible that the rocks this girl is climbing came from a volcano.

Earth's crust

magma

What Is Igneous Rock?

Remember that burning volcano? The liquid rock inside it is called **magma.** Sometimes magma gets squeezed between solid rocks inside the Earth. There, it slowly cools and hardens. Hardened magma is called igneous rock. When something catches on fire it is said to "ignite." Igneous rocks are made from fire, or heat.

Granite is one type of igneous rock. Granite is usually a strong, hard, rough rock with lots of grains in it. It is the perfect kind of rock to climb on!

hot lava, and new rock formed by cooled lava

Lava is magma that gets to the surface of the Earth. When lava cools quickly, it forms rock that has grains of many sizes. This rock also has lots of little holes. It is called basalt.

Look at your three mystery rocks. One of them is basalt. Which one is the right fit?

mountains formed out of granite

What Do Igneous Rocks Look Like?

Granite and basalt are just two types of igneous rock. All igneous rocks share some things in common. They are often dark gray, black, or even pink. Some have holes. They form when hot rock cools. The holes are actually air pockets. Igneous rocks can have grains of different sizes.

Look at the rocks you collected outside. Do any of them look like igneous rocks?

Snow and ice break down the rock on this mountain.

Sedimentary Basics

Sedimentary rocks begin when flowing water, wind, rain, and ice break small rocks off from big rocks and mountains. This breakage is called **erosion.**

Some pieces of rock become **sediments.** They settle in rivers, lakes, and oceans.

How does erosion change the way rocks look?

The water from the river erodes rock, too.

Over time, sediments pile up. New layers push down on old ones. Two layers squeeze into one. This pushing forms sedimentary rocks. Stuff in the ground can get trapped in there. Sometimes these rocks contain gas, oil, or gold.

How might sedimentary rock and igneous rock look different from each other?

Sedimentary rock, such as this piece of rock wall, often forms in layers.

What Are Sedimentary Rocks Like?

Sedimentary rocks are often light in weight. The grains may look glued to each other. Sometimes, you can see layers. Sometimes you can even see fossils. Why might fossils show up in these rocks?

Common rocks in this group are sandstone and shale. Conglomerate is also a type of rock in this group.

Look at your nameless mystery rocks. One is conglomerate. Which one do you think it is?

Some metamorphic rock is pretty. It can even be used for decoration.

What Is Metamorphic Rock?

Metamorphic is the third type of rock. This type of rock starts out as igneous or sedimentary. Then it changes. *Morph* is another word for change. *Metamorphic* comes from a Greek word. It means, "change of form." Metamorphic rocks are not as common as the other two kinds. Why do you think that is?

This slate is a type of metamorphic rock.

Metamorphic rock is made deep below the ground. Another type of rock is under very high heat and very high pressure. Over time, the tremendous heat and pressure change the rock to metamorphic rock.

Do you think any of your mystery rocks come from this group?

Did you know that most chalkboards are made from slate?

What Are Metamorphic Rocks Like?

Metamorphic rocks often have layers. They can have light and dark stripes. Most are hard. Some have large grains. These grains are packed tightly. Common rocks from this group are marble, slate, and schist.

One of your mystery rocks is marble. Which one do you think it is?

23

What Is the Rock Cycle?

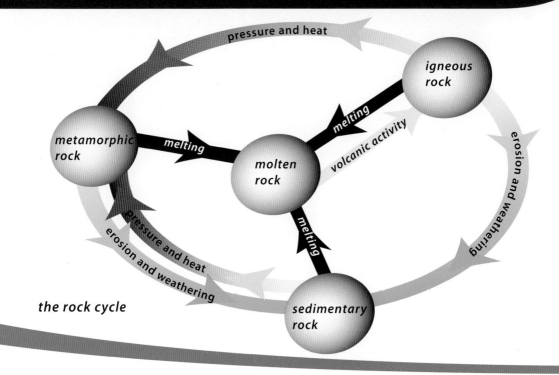

the rock cycle

Rocks feel solid. But they are always changing. It is hot deep inside the Earth. There, rocks melt. They move. They come to the surface and cool down. Old rocks can even change into new rocks. These changes may take millions of years. But rocks never stop breaking down. And they never stop forming again. This breaking down and re-forming is called the rock cycle.

Can you change rocks with your bare hands? What forces have the power to change rocks?

the surface of Mars

Take another look at the rocks you collected outside. Each one has a history. Where is each rock in the rock cycle? Is it breaking down? Or did it just form? Are any of them metamorphic?

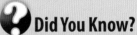 Did You Know?

Some rocks from Mars have little balls of iron in them. Scientists think that water helped form these little lumps of iron. This means there might be water on Mars. If there's water on Mars, there might be life, too!

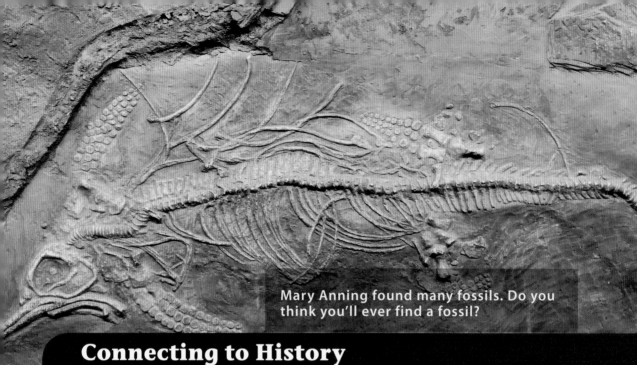

Mary Anning found many fossils. Do you think you'll ever find a fossil?

Connecting to History

Mary Anning

Mary Anning was just a kid. But she was really good at finding fossils. Mary was born in England in 1799. There were lots of sedimentary rocks near her home. As a girl, Mary spent a lot of time looking at the rocks. She was about 12 years old when she and her brother found a skeleton that was almost complete. The animal was a reptile. It lived in the ocean at the same time the dinosaurs lived on the land.

Today, people still know about Mary. She was one of the greatest fossil hunters of all time.

Do you think it's hard or easy to find fossils? Why?

Science at Work

Volcanologist

Volcanologists study volcanoes. They look at minerals, rocks, and lava near volcanoes. They also look at how the Earth moves. They want to know why volcanoes erupt. And they want to know when volcanoes will blow their tops.

Why might it be important to know when volcanic eruptions are going to happen?

This scientist is wearing a mask to protect herself from poisonous gases coming from the volcano behind her.

SOLVE THE iScience PUZZLE

Let's look at your mystery rocks again. Do you know enough now to solve the puzzle? These are the labels that fell off: "Basalt," "Marble," and "Conglomerate."

Rock 1 is dark and has holes. It has both small and large grains. It is basalt, an igneous rock.

Rock 2 has pebbles in it. They look glued together. It also has a fossil in it. It is conglomerate, from the sedimentary group.

Rock 3 has light and dark streaks. It has large grains. It is marble, and is a metamorphic rock.

Were your answers correct when you first read the iScience Puzzle? Did you change your answers by the time you got to this page?

BEYOND THE PUZZLE

Some rocks are much harder than others. Look at the rocks you collected outside. Try to use your fingernail to scratch each one. Next, try to scratch them with a penny. (A penny is harder than your nail.) Now, see which rocks can put scratches on other rocks. Are there any that you can't scratch at all?

Rank your rocks on how hard they are. Use a scale from 1 to 10. Now, look in books or on the Internet. Research the **Mohs scale.** This scale is what experts use to rank hardness in rocks. Look at the scale you made. Is it similar to the one that experts use? How could you make yours better?

Line your rocks up according to how hard you think they are.

GLOSSARY

basalt: a type of igneous rock that forms when lava cools on the surface of the Earth.

conglomerate: a type of sedimentary rock.

erosion: breaking or wearing away of the Earth's surface by wind and water.

fossil: parts, imprint, or trace of a prehistoric animal or plant found in rock or on the Earth's surface.

geologists: scientists who study the Earth.

igneous rocks: rocks that form from molten or liquid rock.

lava: melted rock that erupts from a volcano.

magma: hot liquid found beneath the Earth's surface that cools to form igneous rock.

marble: a type of metamorphic rock.

metamorphic rocks: rocks that form from other rocks.

minerals: natural solid substances that are not made of plants or animals, and that often have flat sides and sharp edges.

Mohs scale: measures the hardness of minerals through the ability of harder materials to scratch them.

properties: things—such as size, color, shape, or texture—that can be observed about an object.

rocks: objects containing one or more minerals.

sedimentary rocks: rocks formed by deposits of sediment.

sediments: bits of material, such as rock and sand, that settle on the bottoms of rivers, lakes, and oceans.

FURTHER READING

101 Things Everyone Should Know About Science, by Dia L. Michels and Nathan Levy. Science, Naturally!, 2009.

Crystal and Gem, by R.F. Symes and R.R. Harding. DK Publishing, 2007.

Rocks for Kids, www.rocksforkids.com.

Children's Museum of Indianapolis, **Geo Mysteries,** http://www.childrensmuseum.org/geomysteries/faqs.html.

ADDITIONAL NOTES

The page references below provide answers to questions asked throughout the book. Questions whose answers will vary are not addressed.

Page 18: Erosion shapes, breaks, and smooths rocks.

Page 19: Sedimentary rock might look layered or like a jumble of rocks glued together. It might contain fossils. Igneous rock might be made of various kinds of minerals, and it may have streaks.

Page 20: Fossils might show up in these rocks because a plant or animal that lies on one layer may be covered by another layer, with its shape preserved.

Page 21: Metamorphic rocks are not as common because they may take longer to form.

Page 24: No, you cannot squeeze rock hard enough to change it. Water, wind, extreme heat, and pressure can change rocks.

Page 26: It is hard to find fossils because most are inside rocks.

Page 27: Volcanic eruptions can be dangerous. If you know when a volcano is going to erupt, you can warn the people living nearby that they must evacuate.

INDEX

Anning, Mary, 26

erosion, 18–19

fossil, 8, 20, 26, 28

geologist, 14

igneous rock, 14–17, 19, 21, 28

lava, 13, 16, 27

magma, 15–16

Mars rock, 25

metamorphic rock, 14, 21–23, 25, 28

mineral, 11–12, 27

properties, 9–10

rock cycle, 24

sediment, 18–19

sedimentary rock, 14, 18–20, 21, 26, 28

volcanologist, 27